Class 73s

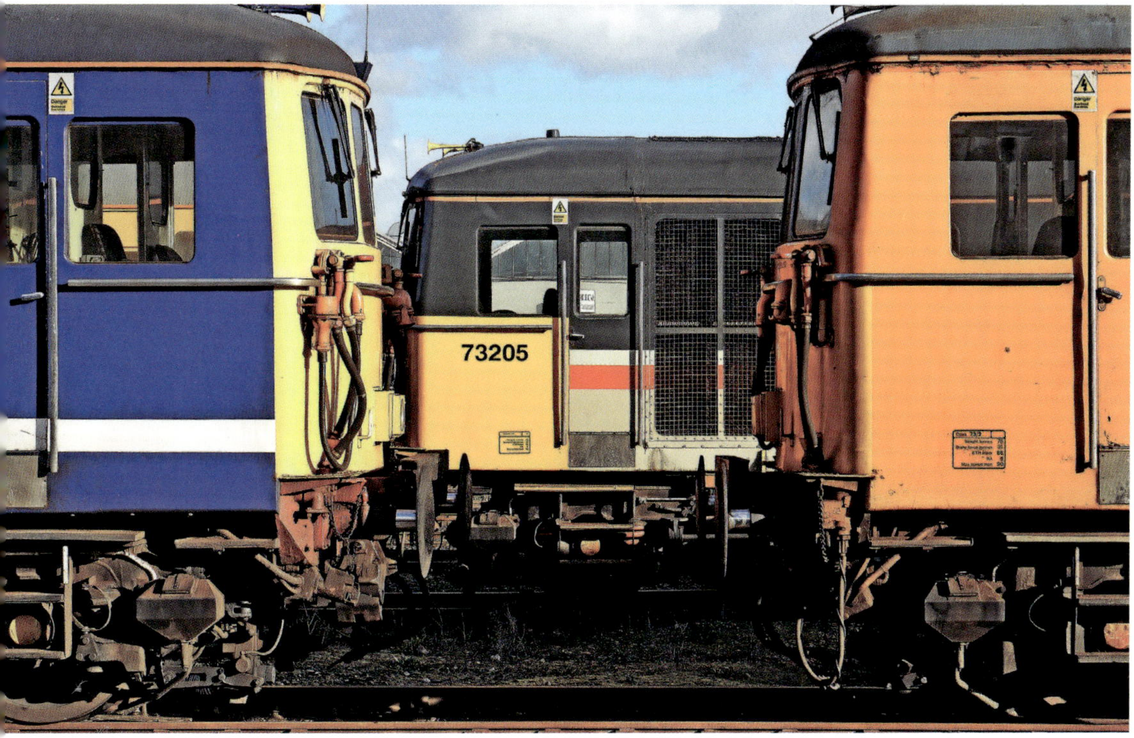

MARK V. PIKE

KEY Books

BRITAIN'S RAILWAYS SERIES, VOLUME 41

Front cover image: This is 73109 approaching Admiralty Bridge, between Holton Heath and Hamworthy, hauling 3-CEP 1497 *Freshwater* and 4-VEP 3417 *Gordon Pettitt* from Weymouth to Bournemouth TRSMD. The very edge of the vast Poole Harbour can just be glimpsed through the trees on the left of picture. 18 October 2008.

Back cover image: 73201 *Broadlands*, 73963 *Janice,* 73136 *Mhairi*, 73141 *Charlotte* and 73212 *Fiona* are seen here running as 0Y68, the 09.39 Tonbridge West Yard to Eastleigh East Yard, passing through Winchester station. 1 September 2021.

Title page image: Three of the class are seen here on the stabling point next to Eastleigh Station. 26 January 2012.

Contents page image: 73107 *Spitfire* is seen coming up the steep incline at Dilton Marsh, soon after leaving Westbury with 1Q10, the 05.12 Westbury to Eastleigh TRSMD via Weymouth test train. 31285 was on the rear of the train doing all the work! 21 April 2010.

Published by Key Books
An imprint of Key Publishing Ltd
PO Box 100
Stamford
Lincs PE9 1XQ

www.keypublishing.com

The right of Mark V. Pike to be identified as the author of this book has been asserted in accordance with the Copyright, Designs and Patents Act 1988 Sections 77 and 78.

Copyright © Mark V. Pike, 2022

ISBN 978 1 80282 544 2

All rights reserved. Reproduction in whole or in part in any form whatsoever or by any means is strictly prohibited without the prior permission of the Publisher.

Typeset by SJmagic DESIGN SERVICES, India.

Contents

Introduction 4

Chapter 1 Class 73/0 5

Chapter 2 Class 73/1 8

Chapter 3 Class 73/2 51

Chapter 4 Class 73/9 81

Introduction

The Class 73s were built between 1962 and 1967, with the first batch of six originally designated JA (later 73/0) and the following production locos being JB (later 73/1). The locos can operate either from a 600hp diesel engine or from 650/750V DC third rail found mostly on the Southern Region. A few were also reclassified 73/2 for use on Gatwick services in the mid-1980s, with some retaining these numbers to the present day.

Following the withdrawal of the ten Class 74 locos, which hadn't proven to be all that reliable, and until the introduction of the Class 88 locos now used by DRS, the Class 73s were unique in their electro-diesel abilities. The original 73s produced a lot more power when on third-rail electric, so they rarely left the Southern Region areas in BR days.

However, much later in life, some were converted to the 73/9 sub-class by GB Railfreight (GBRf), which provided them with brand new engines capable of a much higher rating of 1600hp. Among these 73/9s are six that are now in use on Caledonian Sleeper trains and have no electric function. Probably due to their versatility, many Class 73s are still in service to this day, while six locos have been preserved, and a couple have even come out of preservation to go back into front line service.

This is 73963 *Janice*, 73964 *Jeanette* and 73962 *Dick Mabbutt* in Tonbridge Yard, all attached to Snow and Ice Treatment Trains (SITT). 23 January 2016.

Chapter 1

Class 73/0

The original prototype Class 73/0s were, for me at least, very elusive, and I only managed to grab a few on film. There were only six of them in total.

This is 73003 in the popular BR large logo blue livery entering the sidings next to Eastleigh station, hauling a short rake of coaching stock. Also of note is the BR crew bus going across the bridge in the background, these having long since passed into history, at least for their intended use! The loco, however, is still with us in preservation and can currently be found on the Swindon and Cricklade Railway. 3 June 1992.

Also in large logo and looking very smart, this is 73005 arriving at Woking with the 'Watercress Belle' railtour, which ran as part of the Woking 150 celebrations held this year. In 2015, this loco was completely re-engineered and converted to a Class 73/9. Now numbered 73966, it spends its time in Scotland operating the Caledonian Sleeper trains. Who would have predicted this back then! 29 May 1988.

Back in the days of loco haulage on cross-country services to and from Poole to the north of England, there were occasions when the rostered Class 47 was unavailable due to failure, and rather than cancel the service altogether, any locomotive that was available was commandeered to take the train. Invariably, this was a Class 33, but, on this occasion, it was a very rare appearance of 73006, which had just arrived at Reading with 1E34, the 11.40 Poole to York service. 47580 *County of Essex* can be seen backing onto the other end of the train for the continuation of the journey north. The Class 73/0s were usually restricted to freight trains on the Southern Region at this time, so the authorities must have been desperate on this day! 30 September 1989.

Left: Oh dear, what a state! This is 73002 in Eastleigh East Yard awaiting a tow to Eastleigh Works. Withdrawn by BR in November 1995, it was later preserved at the Dean Forest Railway, acting as spares for 73001, which was also on the railway. However, Locomotive Services acquired both locos during 2019 with a view to returning them to the main line, however, as of October 2022, this has not happened as yet. 2 May 2019.

Below: Later the same day, 73002 was towed to Eastleigh Works by 66779 *Evening Star*. The pair are seen in the yard having just coupled up. This was a 'grab' shot from a carriage window, so it is not the best quality. 2 May 2019.

73003 *Sir Herbert Walker* is seen at Ropley during a visit to the Mid Hants Railway. The loco received BR green livery a few years before it was withdrawn from service and was often used on charter trains, mainly as a back-up or shunt loco for the main train engine. As mentioned earlier, it can now be found on the Swindon and Cricklade Railway. 3 July 1993.

In early 1993, all of the Class 73/0s (apart from 73003) were declared surplus, and in a bit of a surprising move were transferred north to be used on the Merseyrail third rail system for engineering and departmental trains. By 2002, all were again made redundant, with all except 73004 passing into preservation. 73006 eventually found its way to the Severn Valley Railway, but during 2009 it was moved to Eastleigh Works. The loco is seen being dragged through Romsey station by D1015 *Western Champion* to take part in the following weekend's Eastleigh Works centenary celebrations. More recently, 73006 has been rebuilt as 73967 for use in Scotland. 22 May 2009.

Chapter 2
Class 73/1

This is the production run and considered to be the standard locos. Various improvements were made to them, having gained experience from the original Class 73/0 locos, including a more useful top speed of 90mph as opposed to 80mph.

For a while during the late 1980s, Class 73/1s were often used in multiple on the existing stock used on the Weymouth line services while the new Class 442s were being delivered. This is 73106 and 73129 *City of Winchester* propelling a London Waterloo to Weymouth service away from Woking. The Class 73s would only have worked to Bournemouth from where a Class 33/1 would go forward to Weymouth. 73106 was scrapped during 2004, but 73129 is preserved and can now be found at the Cambrian Heritage Railway at Oswestry in fully operational condition and carrying BR 'electric blue' livery. 29 May 1988.

Basingstoke is the location this time as we see 73106 and 73133 waiting to depart with a Weymouth to London Waterloo service. The fate of 73106 was noted in the previous shot, but 73133 is still with us and currently based at Eastleigh Works for use as a shunting loco. Circa 1988.

Brighton Evening Argus

One of the most recognisable of this sub-class was undoubtedly 73101, which was specially repainted in Pullman livery for hauling the VSOE. The loco was first named *Brighton Evening Argus* in 1980 but was re-named *The Royal Alex'* in 1992. It is seen at Wareham just over a month after its repaint at the head of an eastbound track recording train. At the present time, the loco is stored unserviceable at Eastleigh Works awaiting developments. 31 October 1991.

On the rear of the test train seen in the previous shot was 73138 in the so called 'Dutch' yellow and grey Civil Engineers livery. This loco went on to feature on many more test trains in its later career, when it was used by Network Rail and painted in all-over yellow. Unfortunately, it was withdrawn during 2017 and is currently stored at Peak Rail. 31 October 1991.

A few pictures showing the changing appearance of 73133. Carrying the name *The Bluebell Railway*, the loco is shunting slam-door units 1725 and 2203 into the sidings soon after arriving at Wareham with 1Z15, the 09.40 London Victoria to Wareham. Not a lot is known about this charter, but it is believed the participants had the option of travelling by road from here to the nearby Swanage Railway; back then, the connection from the main line was still just a dream. 5 October 1991.

A few years after the previous image was taken, the loco received Network SouthEast livery and it is seen hauling 73108 through Tonbridge station, heading for the large marshalling yard to the west of the station. 73108 was less fortunate than 73133 as it was scrapped during 2004.

73133 later became unique when during the 1990s, it was chosen to be modified for route learning purposes. This involved the headcode box removal from the centre window and the addition of an extra windscreen wiper. The loco also had headlight clusters fitted due to having to compensate for the loss of the headcode box, although unusually these were fitted vertically rather than horizontally so as not to foul the multiple working cables. The loco also received the deep blue Mainline Freight livery. It is seen heading south through Micheldever, hauling a couple of barrier coaches bound for Eastleigh Works. 15 April 2003.

Now in the ownership of Transmart Trains, 73133 was for a while used at Fairwater Yard, Taunton, as a shunter for the High Output Ballast Cleaner trains operated by Freightliner on behalf of Network Rail. It then moved to Bournemouth depot for a while in the 2010s. It is currently on hire to Arlington Fleet Services at Eastleigh Works, and it is seen here undertaking its designated duties, also as a shunter. 4 January 2019.

Affectionately known as 'BoB' by many enthusiasts, 73109 *Battle of Britain* has been a popular loco in recent years. It was taken on by South West Trains in the mid-1990s for use mainly as a 'thunderbird' rescue loco for failed EMUs. The next few images show the loco in action. This one shows it passing Eastleigh hauling Class 423 4-VEP 3482, possibly heading for Bournemouth depot. 27 August 2003.

Left: Next it is captured as it approaches Basingstoke hauling a pair of 4-CIG units bound for Wimbledon depot. With the slam-door units reaching the end of their working lives during the mid-2000s, the loco was kept pretty busy with rescues and stock transfers. 16 February 2004.

Below: Looking absolutely resplendent in its newly applied South West Trains (SWT) 'Desiro' livery, the loco is seen waiting to depart Branksome station on a murky morning with the 5Z50 Bournemouth depot to Wimbledon depot hauling two Class 423 4-VEP units. 1 June 2005.

The last slam-door electric units were withdrawn from the network in late 2005, but South West Trains continued to run a pair on the Lymington branch until May 2010. It had taken two Class 421 3-CIG units and fitted them with central door locking, running the service as a timetabled 'heritage line'. Also retained was BR blue-liveried Class 423 4-VEP 3417 *Gordon Pettitt*, which is now housed at Strawberry Hill depot. This is 73109 approaching Admiralty Bridge, between Holton Heath and Hamworthy, hauling 1497 *Freshwater* and 3417 *Gordon Pettitt* from Weymouth to Bournemouth. The very edge of the vast Poole Harbour can just be glimpsed through the trees on the left of the picture. 18 October 2008.

Right: Coming around the curve from the Bournemouth direction, this is 73109 top and tailing 73235 and arriving at Branksome hauling 3-CIG 1498 *Farringford* while heading for Bournemouth depot. The ensemble will reverse here and veer to the right behind the leading loco to enter the depot. The 'Lymington Flyer' headboard was a local addition that often used to appear around this time. 10 July 2008.

Below: The headboard is seen again as 73109 powers through Winchester hauling 3-CIG 1497 *Freshwater* and heading for Wimbledon depot. 17 May 2007.

Class 73s were often associated with workings to/from the Atomic Energy Authority establishment at Winfrith, near Wool. These were often lightly loaded and were within the haulage capabilities of the diesel engine in the loco. This was required as they often worked as far as Westbury, which was well off the third rail electrification area. This is 73119 *Kentish Mercury* leaning into the curve through Hamworthy station with 7Z96, the 10.33 Winfrith to Gloucester, which would be taken forward from Westbury later in the day, usually by a Railfreight Coal sector Class 31, with the 73 returning back to the Southern Region. This loco is still very much with us today, earning its keep on the main line operated by GBRf. 10 April 1990.

On another occasion, the train was hauled by 73140 and is seen this time passing Eastleigh. This loco has since been preserved, and the immaculate loco can now be found on the Spa Valley Railway at Tunbridge Wells. 9 October 1990.

Back in BR days, it was unusual to see any of the class away from third rail territory, but as mentioned previously, this flask working did get them as far as Westbury, just about! The 600hp engine was sometimes prone to overheating when used for prolonged periods. This is 73117 *University of Surrey* just arrived at Westbury, where it was removed in favour of 31276 *Calder Hall Power Station* for the continuation of the journey. The 73 has now become 73970 and works north of the border on Caledonian Sleeper services. 19 March 1991.

Another source of work for the locos has been with charter trains over the years. This is 73101 *The Royal Alex'* and 73131 arriving at Wareham with the empty stock from 1Z86, the 09.00 London Victoria to Poole 'The Wessex Venturer' organised by Hertfordshire Railtours. The train stabled at Wareham until later in the day. 73101 is currently stored at Eastleigh Works while 73131, one of only two of the class to receive EWS maroon livery, was scrapped during 2004. 28 April 2001.

Here we see 73110 and 73136 *Kent Youth Music* just arrived at Bournemouth with 1Z82, the 09.00 London Victoria to Bournemouth 'The Dorset Ooser', which was organised by Hertfordshire Railtours. The Ooser is part of 18th/19th century Dorset folk culture. They were rather scary and menacing masks used in traditional midwinter and May Day gatherings. The purpose was either to light-heartedly scare people in village revels, or to humiliate those who were deemed to have behaved immorally! Anyway, 73110 is currently stored unserviceable at Eastleigh Works while 73136 is in main line service with GBRf. 20 October 2001.

During the mid-2000s, one of the most well-known locomotive providers was Fragonset, which used a distinctive black livery for its locos. This is 73107 *Spitfire* 'on the blocks' at London Waterloo, having brought in the empty stock for 1Z92, the 10.00 Waterloo to Salisbury 'Cathedrals Express' which was hauled by 'Black Five' 45231. The Class 73 was the only one of its class to receive the all-black livery. It is now in standard GBRf livery and sees regular use on the main line network. 12 December 2006.

The loco is seen again at Salisbury on the rear of 1Z92. The loco remained attached purely for any shunt moves that were required. This would have been another unusual location to see a Class 73 in BR days. 12 December 2006.

The ever popular 73109 *Battle of Britain* is seen in immaculate condition as it awaits departure with 3-CIG 1498 *Farringford* as 1Z43, the 14.20 London Waterloo to Brockenhurst staff charter. This train was organised as a surprise for an employee from the South West Trains control office at Waterloo; it was his last day at work after 43 years on the railway. Note that headboard again! 10 July 2008.

Right at the other end of the line, another immaculate loco awaits departure from Weymouth. This is 73136 *Perseverance* in early BR blue livery with the 1Z83 Weymouth to London Waterloo 'Dorset Dumpling' charter organised by Spitfire Railtours. 33025 *Glen Falloch* was on the other end of the train and had brought it down from London earlier in the day. 6 September 2008.

Emerging from the mist and murk, 73141 *Charlotte* and 73204 *Janice* are approaching Dorchester South with 1Z92, the 08.40 London Waterloo to Weymouth 'The Weymouth Seaside Special' organised by UK Railtours. 73141 was one of a few of the class to receive the rather attractive deep blue GBRf livery in the late 2000s, but this example also had an oversized number on the cabsides; the loco still operates on the main line to this day with GBRf. 73204 was one of the locos to be chosen for rebuilding as a Class 73/9 in the mid-2010s and eventually became 73962 *Dick Mabbutt*. 1 August 2009.

With covered nameplate, yet another immaculate machine as 73136 is seen six years on from the previous image (with 73107 on the rear) and captured passing through Farnborough with the 5Z69 Eastleigh to London Victoria empty stock for a private charter later in the day. Upon arrival in London, both locos were named; 73107 *Tracy* and 73136 *Mhairi*. 25 August 2015.

Possibly one of the most audacious railtours to have run in recent years is this one! 73128 *OVS Bulleid CBE* and 73107 *Tracy* splutter past Sprey Point on the approach to Teignmouth with 1Z73, the 08.00 London Waterloo to Paignton 'The Herd of Wildebeest' charter. On the rear of the train can just be seen 73962 *Dick Mabbutt* and 73963 *Janice*, which had led the train for the bulk of the journey from London as far as Exeter St Davids, from where the two 73/1s took over for the last leg to Paignton. However, the heavy train proved a bit much for them as they both failed not long after the shot was taken between Newton Abbot and Torquay and were pushed by the rear locos the final few miles to terminate at Paignton around 55 minutes late. 16 July 2016.

A few years later and 73107 *Tracy* was once again off region and seen at Westbury on the rear of 5Z83, the 11.25 Warminster to Heywood Road Junction (to turn the whole train) hauled by Swanage Railway-based Class 33 D6515 *Lt Jenny Lewis RN*. The train had terminated at nearby Warminster so passengers could alight to board a fleet of Routemaster buses that operated from Warminster to the nearby lost village of Imber on Salisbury Plain, which was taken over by the MoD during wartime and is usually out of bounds to the public. 17 August 2019.

The first of a few images taken when I was working on the railway many moons ago! It is hard to believe that this picture was taken from the former down platform at Boscombe station, just east of Bournemouth. 73103 is just getting into its stride having departed Bournemouth around five minutes earlier and hauling two 4-TC units heading for London Waterloo. Boscombe was quite a substantial station but was closed in October 1965 just prior to electrification of the route. The loco itself is now working at the opposite end of the UK, having been converted to 73968 during 2016 for use on the Caledonian Sleeper trains. 16 February 1988.

Pulling out of No. 2 middle siding at Bournemouth Central, this is 73123 *Gatwick Express* hauling two 4-TC units that will form a service to London Waterloo. Around this time. the powerful Class 430 4-REP electric units were gradually being withdrawn to provide parts for the new Class 442 units, so Class 73s were used as a temporary replacement. This loco was completely re-engineered during 2014 and is now operating for GBRf as 73963 *Janice*. 17 February 1988.

In the popular large logo livery, 73114 has just exited the short tunnel at Holdenhurst Road upon departure from Bournemouth Central with another service for London Waterloo. Note the inclusion of a buffet car in the first unit, just about a thing of the past these days! This loco is currently stored at Nemesis Rail, Burton upon Trent.

A little further east from the last image we see 73138 *Post Haste* and 73135 passing beneath Gloucester Road bridge near Boscombe with another London Waterloo service. Although 73138 is in InterCity livery, it is unusual in having the full yellow ends extending around the window frames. It is now stored at Peak Rail. 73135 later went on to be re-numbered 73235 to work with Gatwick Express but can now be found pottering around as the Bournemouth depot shunting loco. 4 October 1988.

The classic location of Bournemouth Central is the location for 73137 *Royal Observer Corps*, always recognisable due to its RAF blue-backed nameplates, as it awaits departure for London Waterloo. Unusually, the front unit is missing its driving coach behind the loco in this instance. 17 February 1988.

Bournemouth depot is situated around three miles or so from Bournemouth Central and is accessed via a junction at Branksome station. 73124 *London Chamber of Commerce* has just come off the depot and the driver is changing ends before taking the train to Bournemouth to form another service to Waterloo. This loco was another one completely re-engineered during 2014 and is now operating for GBRf as 73964 *Jeanette*. 17 February 1988.

This is 73110 and 73105 *Quadrant* powering up the 1 in 50 gradient soon after leaving Poole with a London Waterloo train. 73110 is currently stored at Eastleigh, while 73105 was converted to 73969 in 2016 for use on Caledonian Sleeper trains in Scotland. Circa 1988.

Class 73s have frequently been used on various test trains throughout the south of England, especially during the 2000s and 2010s. The next set of pictures shows some of these workings. This is a remarkable shot in which no less the six of the class can be seen! 73109 and 73138 are seen at Eastleigh with 1Q07, the 06.05 Hither Green to Eastleigh Works, while in the background are 73212, 73119, 73141 and one unidentified. 23 October 2013.

Left: The time on the Civic Centre clock is 08.34, and the sun is not long risen. Captured at the split second that 73141 *Charlotte* lights up the morning with a huge arc on the third rail, this is 1Q38, the 06.08 Woking to Woking test train passing through Southampton Central. 5 January 2016.

Below: With the small uninhabited Pergin's Island above the train, this is the fine view across a small part of the vast Poole Harbour, as 73107 *Redhill 1844–1994* and 73138 come across Holes Bay causeway near Poole with 1Q08, the 05.19 Eastleigh Works to Eastleigh Works via Fareham, Weymouth and Portsmouth. 19 April 2011.

Another two-tone grey example around the same time was 73128 *OVS Bulleid CBE*, and it is seen arriving at London Waterloo with 1Q82, the 12.03 Woking Up Carriage Sidings to Woking Up Carriage Sidings test train. 24 February 2015.

Attached to the other end of the train seen in the previous image was 73136, and it awaits departure from beneath the vast glass roof of London Waterloo station. 24 February 2015.

Above: Bright yellow Network Rail-liveried 73138 (with 73109 on the rear) is seen approaching Fareham with 1Q08, the 05.06 Eastleigh Works to Eastleigh Works via Fareham, Eastleigh, Weymouth, Fratton and Southampton Central. 24 October 2013.

Left: Passing through the heavily canted centre road at Farnborough, this is 73141 *Charlotte* leading 1Q82, the 13.49 London Waterloo to Southampton test train. 73212 was on the rear. 26 January 2010.

73107 *Redhill 1844–1994* is seen crossing the picturesque Eynsford Viaduct, near Swanley, Kent, with a Hither Green to Hither Green test train. 29 September 2011.

Above: 73141 *Charlotte* with 73201 *Broadlands* on the rear pass through Staines Down Goods Loop with 1Q53, the 08.51 Hither Green to Woking test train. 11 February 2015.

Right: With an aircraft coming in to land at the nearby Southampton Airport on a cold winter's morning, we see 73138 leading 1Q41, the 08.02 Eastleigh Works to Eastleigh Works via Portsmouth and Southsea and MoD Ludgershall test train, as it crosses the viaduct over the River Itchen near St Denys. 73107 was on the rear. 14 January 2014.

Soon after being taken on by GBRf, this is 73109 *Battle of Britain* bringing up the rear of an unidentified test train at Fratton. The train had just come the short distance from Portsmouth Harbour and was probably eventually destined for Eastleigh Works. The patch below the nameplate erased the name of the former operator, South West Trains. 21 June 2013.

By the end of 2013, the former South West Trains livery carried by 73109 had been removed completely, and for a while the loco carried this plain light blue base colour with no nameplates. The loco is seen approaching St Denys on the rear of 1Q08, the 05.06 Eastleigh Works to Eastleigh Works. 24 October 2013.

73107 *Redhill 1844–1994* was partnered with 73201 *Broadlands* on the 1Q58 Hither Green to Hither Green test train, just after passing through Hollingbourne, Kent. 23 August 2012.

A location that has since changed out of all recognition, this is 73141 *Charlotte* passing through London Bridge station with the 1Q66 Charing Cross to Charing Cross via Hastings circular test train. Since the date of this image, the station has been completely rebuilt and a certain tall building called The Shard now dominates the background. 27 November 2009.

This is 73141 *Charlotte* leading 1Q82, the 05.47 Selhurst to Waterloo via Weymouth test train, and heading west through Poole station. 26 January 2010.

Above: Just emerged from Eastleigh Works on a bitterly cold morning, this is 73119 *Borough of Eastleigh* with 1Q13, the 09.18 Eastleigh Works to Tonbridge West Yard via Fareham and the Sussex coastway line. 73206 on the other end would lead the train from here. 16 January 2013.

Left: Unusually operating on the same day as the one in the previous picture, this is 73107 *Redhill 1844–1994* top and tailing with 73201 *Broadlands* on 1Q43, the 06.31 Eastleigh Works to Eastleigh Works via Hamworthy Goods and Weymouth, crossing the causeway at Redbridge, just west of Southampton. 16 January 2013.

Apart from the regular third rail timetabled passenger service with EMUs, the spur from the Staines line that comes into Weybridge on the South Western Main Line very rarely sees any other type of train, so to see a loco-hauled train is exceptional. However, this is 73107 *Redhill 1844–1994* approaching Weybridge on this spur with an unidentified test train. 73138 was on the rear. 20 January 2011.

Once again, this is 73107 *Redhill 1844–1994*, this time unusually paired up with 73201 *Broadlands* and 73138 on the rear of 1Q43, the 06.30 Eastleigh Works to Eastleigh Works, via Hamworthy and Weymouth test train. 11 January 2012.

Right: The train seen in the previous image returned later the same morning and is captured crossing Holes Bay causeway on the approach to Poole with the pair now leading. It is not known as to why there were two locos on one end of this train, but quite possibly one was not performing as expected? 11 January 2012.

Below: Rather surprisingly, a couple of weeks after the last images were taken, the three locos were still together on another test train! This time we see 73138 leading with 73107 *Redhill 1844–1994* and 73201 *Broadlands* on the rear as 1Q45, the 12.24 London Bridge to Hither Green Down Reception, passes Selhurst depot. 27 January 2012.

A quick leap across the road sees 73107 *Redhill 1844–1994* and 73201 *Broadlands* on the rear of the train as they approach Norwood Junction station. 27 January 2012.

Now for a few images of loco convoys. This is an Eastleigh to Tonbridge Yard working passing through Winchester with 73136 *Perseverance*, 73209 *Alison*, 73205 *Jeanette*, 66703 *Doncaster PSB* and 66705 *Golden Jubilee*. Since the rebuilding of the bridge I took this from, it has been very difficult to photograph here with higher parapets. 31 October 2008.

About three or four miles further south at Shawford, we see 73109 *Battle of Britain*, 73136 *Perseverance*, 73204 *Janice* and 73206 *Lisa* passing through as 0Y66, the 09.30 Eastleigh Yard to Tonbridge West Yard. 14 March 2012.

Taken in very dark conditions, this is 73141 *Charlotte*, 73208 *Kirsten*, 73213, 73206 *Lisa* and 73212 approaching Winchfield and running as 0Y68, the 09.30 Tonbridge Yard to Eastleigh Works, which was over an hour late at this point. 10 December 2010.

Formed up in Eastleigh Works and ready to go, this is 73136 *Perseverance*, 73211, 73210, 73204 *Janice* and 73205 *Jeanette* with 0Z73, the 12.10 Eastleigh Works to Tonbridge West Yard. All the Class 73/2s in this convoy were later converted to 73/9s, with the exception of 73210, which was preserved and can now be found on the Ecclesbourne Valley Railway. 73136 is still in service with GBRf. 26 October 2005.

A shot that is not quite all that it seems! 73141 and 47739 are actually being shunted into the works complex at Eastleigh unusually by preserved Class 52 D1015 *Western Champion*. The 73 and 47 were later refurbished for GBRf and Colas Railfreight, respectively. In the last few years, however, 47739 has transferred to GBRf mainly for use in dragging various EMUs off their normal operating regions. 15 November 2007.

The next images show locos off their normal operating area. Although it is not so rare to see a Class 73 'off region' nowadays, back in BR days it was pretty unusual. One morning in the late 1980s, I got off a train at Swindon, deep in Western Region territory, and was very surprised to say the least to see 73135 just pulling away. Luckily, though by no means perfect, I was just able to grab this shot. It was presumably heading back to more familiar surroundings in third rail land? I still have no idea to this day the reason for its visit. Circa 1987/88.

At the time I took the last image, a working such as this would have been unthinkable! 73107 *Spitfire* is seen coming up the steep incline at Dilton Marsh, soon after leaving Westbury with 1Q10, the 05.12 Westbury to Eastleigh depot via Weymouth. Interestingly, the Class 73 was actually being used as a driving trailer and not under power; the loco providing that power being 31233, which was giving its all on the rear! 21 April 2010.

Taken a day before the last image, this shot is very much within the operating region of Class 73s, but only a handful have ever made it to here, mostly on test trains. 73107 *Spitfire* stands at the non-electrified Hamworthy Goods, at the end of the two-mile-long route from Hamworthy Junction with 1Q08, the 09.28 Bournemouth depot to Westbury via Hamworthy and Fratton. Once again, 31233 on the other end was doing most of the donkey work. 20 April 2010.

As noted earlier, Class 73s were occasionally drafted in at short notice to work the loco-hauled cross-country services from Poole if a rostered Class 47 was not available for any reason. The Class 73/1 variant was more commonly seen simply because there were many more of them than 73/0s, but still pretty rare, nonetheless. Amazingly though, a couple appeared within just two of months of one another during 1991. Firstly, we see good old 'BoB', 73109 *Battle of Britain*, being detached from 1M88, the 06.40 Poole to Manchester Piccadilly at Reading. The loco returned back south almost straight away, and the train was taken forward by a Class 47. 14 June 1991.

Long before its second lease of life on the main line with GBRf, this is 73136 in plain grey livery arriving at the same platform at Reading as in the one in the previous shot with an unidentified service from Poole to the north of England. Once again, the loco promptly returned back to its native Southern Region soon after being detached from the train. One enthusiast seems to be enjoying the unusual haulage! 14 August 1991.

The first of a couple of images from the celebrations held in 2008 to commemorate the 150 years of the Lymington branch in Hampshire. To mark the occasion, 73109 *Battle of Britain* was used to haul 3-CIG 1498 *Farringford* on timetabled services throughout the day. The appearance of any sort of loco has always been very rare on this five-mile branch and still is to this day. Here, we see the loco arriving in to Brockenhurst with 1J12, the 08.44 Lymington Pier to Brockenhurst. The SWT 'Desiro' livery of the loco is a nice contrast with the BR green unit. 12 July 2008.

A little later in the morning we see the same combination coming across a barren part of the New Forest National Park, near Setley, with 1J18, the 10.27 Lymington Pier to Brockenhurst service. On the skyline can be seen the local landmark of Sway Tower, which was built during the 1880s, simply because the landowner was a wealthy person and he wanted a centrepiece for his estate! 12 July 2008.

Now to the first of a batch of images taken of Class 73/1s that have visited the Swanage Railway in Dorset, mainly for the annual diesel galas. Quite often, a 73 was more or less just used as a brake translator for the various types of other locos that could not operate solo with the Swanage Railway coaching stock. However, some have got to do a few trips solo. This is 73119 *Borough of Eastleigh* spluttering away as it comes beneath Townsend Bridge on Corfe Common, making its way to Swanage hauling 4-VEP 3417 *Gordon Pettitt*. 9 May 2010.

The same loco and unit are seen the previous day passing through Harmans Cross station with a Swanage to Norden train. The headcode '90' used to be carried by the Weymouth Quay to London Waterloo boat train that ran during the 1970s and 1980s. 8 May 2010.

A few years earlier, 73136 *Perseverance* visited the line as well as 4-VEP 3417 *Gordon Pettitt*. Dating back to the 11th century, the spectacular ruin of Corfe Castle dominates the backdrop, as the formation is captured crossing Corfe Common with Class 33 D6515 bringing up the rear. 11 May 2007.

The castle ruins are dominant again as we see the paring of 73136 *Perseverance* and 73208 *Kirsten* awaiting departure from Corfe Castle with a Swanage to Norden train. 73208 was later converted to 73965. 12 May 2007.

Now in green, 73136 made another visit to the line during 2012. It is seen coming round the curve at Afflington on the approach to Harmans Cross with a Norden to Swanage train. The green livery of the loco matched the Swanage Railway Mk.1 coaches very well. 13 May 2012.

Also putting in an appearance the previous year, the colourful pairing of 73136 *Perseverance* and 73205 *Jeanette* are approaching Corfe Castle with a Norden to Swanage train. 73205 later went on to become 73964. 6 May 2011.

Ten years after its last visit, and now in standard GBRf livery, 73136 *Mhairi* was back again for another diesel gala, although it did not actually head that many trains. It is seen with Swanage home loco 33111 approaching Corfe Castle with a Swanage to Norden train. The headcode '98' was the correct one for Swanage branch trains in BR days. 6 May 2022.

Not exactly an everyday location for a Class 73! This is 73136 *Mhairi* resting between duties on Swanage shed, along with the distinguished company of ex-BR Battle of Britain 'Pacific' 34072 *257 Squadron* and T9 30120 peeping out of the shed. 6 May 2022.

During the 2019 diesel gala, 59003 *Yeoman Highlander* was an unusual visitor but was unable to work with the Swanage stock on its own. This time it was the turn of 73119 *Borough of Eastleigh*, also now in standard GBRf livery, to act as the brake translator. The pair are seen departing from Harmans Cross with a Swanage to Norden train. 11 May 2019.

Left: If I remember correctly, there was only one working for 73119 *Borough of Eastleigh* during the gala in 2019 and this is it, seen crossing Corfe Common with an afternoon train from Norden to Swanage. 11 May 2019.

Below: The final view on the Swanage line in this section sees immaculate 73107 *Spitfire* departing Swanage hauling 4-VEP 3417 *Gordon Pettitt* as they head for Norden. 9 May 2008.

Not too far from the Swanage Railway is Wareham station. Here, we see 73109 *Battle of Britain* in Network SouthEast livery awaiting a green signal while hauling a 4-VEP as a crew training special from Weymouth to Bournemouth. 17 March 1995.

Having been given a favourable signal, the train pulls away from Wareham and across the notorious foot crossing. The crossing was once classed as one of the most dangerous in the country after many near misses with pedestrians and trains, so much so that in the late 2000s a permanent attendant was posted to assist users. It had been earmarked for closure, but it is a very important crossing point, the only other way across the line is via the flyover (with no walk ways) that was built to replace the crossing back in the 1980s. Most of these recent problems could probably have been avoided had it remained as a road crossing as it was many years ago! The signal box above the loco was closed in May 2014 but has been kept intact for occasional further use by the Swanage Railway as and when required. 17 March 1995.

Along with the Multi-Purpose Vehicles (MPVs), another duty often entrusted to the small pool of Class 73s operated by GBRf are the annual Railhead Treatment Trains (RHTT) in the south of England that have operated from the mid-2010s. This is 73128 *OVS Bulleid CBE* passing Kensington Olympia bringing up the rear of 3W90, the 04.30 Tonbridge West Yard to Tonbridge West Yard via many lines around the London area. 73207 was leading the train. 17 November 2014.

Left: A few years later, and the same 3W90 04.30 Tonbridge West Yard to Tonbridge West Yard is again seen at Kensington Olympia, this time headed by 73136 *Perseverance* with 73201 *Broadlands* on the rear. 10 October 2018.

Below: This time 3W90 is seen passing through platform 17 at Clapham Junction lead by 73119 *Borough of Eastleigh* with 73141 *Charlotte* bringing up the rear. 9 December 2016.

Occasionally, the specialised RHTT vehicles are tripped to Eastleigh Works for maintenance and are often hauled by a Class 73. This is 73136 *Mhairi* caked in the mess that arises from constant rail spraying with a few wagons approaching Basingstoke with 3W35, the 11.10 Tonbridge West Yard to Eastleigh Works. 4 December 2015.

Heading in the opposite direction a couple of years later is 73109 (then un-named) passing Hook station with 3Y82, the 12.47 Eastleigh to Tonbridge Yard with a couple of wagons off maintenance. 23 November 2017.

One of the most recent innovations for keeping the third rail clear during the winter months are the Snow and Ice Treatment Trains (SITTs). These wagons are also loco-hauled and work alongside the MPVs again. This is Tonbridge station with 73119 *Borough of Eastleigh* passing through having just exited the nearby yard with 3Y01, the 11.05 Tonbridge Yard to Tonbridge Yard circular SITT train on a thoroughly miserable day. 23 January 2016.

To finish this section on Class 73/1s, we have a series of images showing general workings for the locos. This first one sees 73114 and 73138 waiting in the platform at Eastleigh for a signal to proceed to the East Yard with a train of sand. 73114 is now owned by Nemesis Rail and is currently stored at Burton upon Trent, while 73138 went on to see much work with Network Rail on test trains until it was stored during 2017. 20 June 1991.

During the mid-1980s, one loco that stood out from the crowd, even from a distance, was 73118 *The Romney, Hythe and Dymchurch Railway* with that massive nameplate! The loco is seen stabled at Basingstoke during the 1987 Rail Fair. This loco later went on to operate with Eurostar and was fitted with Scharfenberg coupling equipment to enable it to haul Class 373 units. More recently, the loco has been preserved and can now be found at the Barry Tourist Railway in South Wales. 26 September 1987.

Throughout the 1970s and 1980s, it would not have been unusual to see a loco stabled in the up bay platform at Bournemouth, but as the years passed it has become far more uncommon. This is 73107 *Spitfire* in its all-over black livery of then operator Fragonset Railways. The loco was here in conjunction with a railtour in the area. 31 May 2005.

Another loco to have carried a large nameplate (although not on the scale of 73118) was 73134 *Woking Homes 1885-1985*, seen here on display at the Eastleigh open day. Quite literally as I write this caption, the loco has just been cut-up in September 2022. 12 October 1986.

Looking quite smart in its Network SouthEast livery, this is 73129 *City of Winchester* shunting around the stabling point next to Eastleigh station. This loco is now preserved in BR 'electric blue' livery and can be found on the Cambrian Heritage Railway. 16 December 1992.

This is preserved E6047 (73140) in the aforementioned early BR blue livery moving around the yard at Tunbridge Wells West which is its home base on the Spa Valley Railway. Just recently, the loco has received a new coat of BR standard blue livery and has been re-numbered back to 73140. 5 August 2006.

With a front-end modification that did absolutely zero for the locos' appearance, Eurostar-operated 73118 and 73130 pass through Kensington Olympia heading south. As mentioned earlier, the cumbersome-looking attachment on the front of the locos was a Scharfenberg coupler to enable them to work with Class 373 Eurostar units in case of failure. However, I mostly remember the locos just running around light engine or parked up in sidings most of the time! Both locos survive in preservation, 73118 at the Barry Tourist Railway and 73130 at MoD Bicester. Circa 1996.

Right: A dismal winter morning sees 73107 *Spitfire* passing Basingstoke with 5Z73, the 09.40 Eastleigh Works to Woking and return Network Rail driver training run. A few years prior to this, one of these runs would have probably been made using a slam-door EMU, but as these had been withdrawn by this date a rake of coaches was mustered up. 15 December 2009.

Below: A very unusual working this time. 73136 and 73119 *Borough of Eastleigh* are seen passing East Croydon with 5O73, the 05.25 West Ruislip London Underground depot to East Grinstead Sidings. This was conveying the superb ex-Metropolitan Railway Chesham Set for use on the Bluebell Railway. 27 August 2014.

This one is even more unusual though! 73119 *Borough of Eastleigh* and 73128 *OVS Bulleid CBE* are hauling London Underground inspection saloon LT45029 and Metropolitan Railway electric loco No. 12 *Sarah Siddons* as 2Z82, the 10.45 Eastleigh Works to West Ruislip depot. Due to an axle box issue on the vintage electric loco, the train was delayed in the Wallers Ash area for quite some time until it was deemed safe to continue. 16 May 2019.

In the early morning spring sunshine, 73119 *Borough of Eastleigh* and 73213 *Rhodalyn* are approaching Basingstoke with a 6G27 05.25 Lingfield to Eastleigh Yard empty long-welded rail train. 28 April 2016.

Yet another unusual working. 73141 *Charlotte* and 73206 *Lisa* with 73205 *Jeanette* and 73119 *Borough of Eastleigh* on the rear pass through Southampton Central with 7Y08, the 07.49 Eastleigh Works to Southampton Western Docks hauling ex-South Eastern 508212 sandwiched between a couple of barrier coaches. The withdrawn unit was being taken to Southampton Docks for onward road movement to Merseyside and further use as a training unit for the fire service. 14 August 2012.

Civil Engineers 'Dutch'-liveried 73128 *OVS Bulleid CBE* is captured while shunting in the yard at Eastleigh. The unique nameplates fitted to this loco are in the style of the ones carried by ex-Southern Railway 'West Country' Class locomotives, which were designed by OVS Bulleid. The loco still carries these plates to this day. 29 October 1991.

A year later, 73128 *OVS Bulleid CBE* is seen again as it passes through Basingstoke with a Hoo Junction to Eastleigh Yard engineers' train. 10 July 1992.

The last image for this section sees 73136 *Perseverance* approaching Basingstoke with 6G15, the 11.15 Eastleigh to Kew Bridge weekend engineers' train. This loco did not work that many engineers' trains while wearing this livery of BR early blue. 16 December 2006.

Chapter 3
Class 73/2

This sub-class was first created back in the late 1980s when a batch was dedicated to working the Gatwick Express services out of London Victoria. Initially, their diesel engines were isolated, but later in their careers, upon cessation of these services, the engines were once again activated. Since the mid-2000s, there has not really been much of a distinction between these and Class 73/1s, the only difference really being the numbering. In the mid-2010s, however, most 73/2s were actually substantially rebuilt, becoming Class 73/9s, which we will see later in the book.

At the end of March 1988, Bournemouth depot held its first ever open day to celebrate the introduction of the brand new Class 442 Wessex Electric units, and amongst the variety of visiting units and locos we see an immaculate InterCity-liveried 73201 *Broadlands* on display. This loco still operates to this day with GBRf, now returned to BR blue livery. 26 March 1988.

During 1989, Class 33 33115 was withdrawn from service with a seized-up engine, but instead of being scrapped it was converted into a Driving Van Trailer (DVT) in connection with the testing of the new Channel Tunnel trains then being developed. After some cosmetic restoration, it was released during 1990 in InterCity livery and carrying the number 83301. The loco had been fitted with new bogies of the design to be used under the Eurostars to test their third rail pick-up shoes. The reason for its original withdrawal, the seized engine, was retained as a dead weight. 73205 *London Chamber of Commerce*, part of the pool of locomotives used for the Gatwick Express, was chosen to pair with 83301, and they were semi-permanently coupled together. This now enabled engineers to test the combination of the new bogies and third rail pick-up gear at speeds of up to 90mph. Strangely, although 73205 actually provided the haulage, the power to drive it from the third rail came through the pick-up shoes fitted to 83301. Recently redundant Class 438 4-TC 8007 completed the formation, and much of the testing was carried out on the South Western Main Line until 1994. The train is seen just after arrival in the sidings at Eastleigh. 20 June 1991.

The train is seen at Eastleigh again at a time when the testing had almost been completed. The former Class 33 was later donated to the Class 33/1 Preservation Group and was used as a source of spare parts for their other locos. The 73/2 has gone on to see much service over the subsequent years with GBRf and during the mid-2010s was given a new lease of life, being rebuilt as 73964 *Jeanette*. 2 March 1994.

The first of a few views of Gatwick Express services now as we see 73210 *Selhurst* propelling a Gatwick Airport to London Victoria train through Purley station. This loco has since been preserved and can currently be found at the Ecclesbourne Valley Railway in Derbyshire, although it is currently not operational. 1 December 1992.

Heading in the opposite direction with a London Victoria to Gatwick Airport train is 73212 *Airtour Suisse*. At the time, none of these services stopped at any of the intermediate stations. This loco is still in main line service with GBRf and named *Fiona*. It is also the oldest 73 in everyday use on the UK national network. 1 December 1992.

This time it is 73235 heading south through Purley with another London Victoria to Gatwick Airport train. This loco (ex-73135) was added to the Gatwick Express pool sometime after most of the original fleet and was initially intended as cover for 73205 while it was engaged on its test train duties but ended up remaining on the services. This loco is still operating today, not on the main line, but as a depot pilot for South Western Railway at Bournemouth. 1 December 1992.

The last view at Purley sees 73208 propelling a London Victoria service almost at the very end of Class 73 haulage, this loco being one of the last to work them. Another example to have seen many years of work in the following period with GBRf, it was converted to 73965 during the mid-2010s and currently carries the name *Des O'Brien*. 10 February 2004.

The last Gatwick Express shot sees 73208 again hauling the usual eight Mk.2 air-conditioned coaches with a Class 489 Gatwick Luggage Van (GLV) on the rear, this time passing through Horley station not far from its destination at Gatwick Airport. 10 February 2004.

When the 73/2s were working in the Gatwick Express pool, they were rarely seen doing anything else as they were used quite intensively. However, a couple were used (though not with the Gatwick Express stock) on 'shoppers' charter trains that ran on two weekends at the end of 1992 from London Victoria to Weymouth and return. This is 73204 *Stewarts Lane 1985-1998* coming across Holes Bay, on the approach to Poole, with the return train to London. 12 December 1992.

Not long after it was made redundant from Gatwick Express services, 73208 was taken on by GBRf and for some time after, it was still carrying its InterCity-derived livery. It is seen passing through Havant in the company of sister loco 73209 *Alison* with an engineers' train bound for Eastleigh Yard. 20 January 2006.

Two ex-Gatwick locos, 73212 and 73213, were used for a while by Network Rail after withdrawal of the class from these trains. They were painted in this very bright all-over yellow livery in keeping with most of the company's fleet of vehicles at the time and used on test trains for a couple of years or so. The pair are seen stabled in the sidings at Eastleigh. 6 December 2007.

This time we see 73212 and 73204 *Alison* approaching Vauxhall station near the end of their journey with the 1Q85 London Waterloo to London Waterloo test train that ran via the Hamworthy Goods freight line. Notice the blue smoke at the rear of the train caused by arcing on the third rail. 26 February 2010.

The next few shots show various EMU 'drags' that have utilised the class, often as training/route learning runs for SWT drivers. Wearing one of the best liveries applied to the Class 73s in my opinion, SWT-operated 73201 is seen propelling EMUs 1497 *Freshwater* and 3417 *Gordon Pettitt* through Poole station and into the sidings. During its time with South West Trains, the loco did not carry its *Broadlands* nameplates. 26 August 2005.

By the time the ensemble left Poole, I had been able to get this second view as it briefly awaits a green signal at Christchurch station to proceed. 26 August 2005.

The loco has now run round the train, and I was able to obtain this final view as it returns south on the approach to Beaulieu Road station in the New Forest National Park. 26 August 2005.

73235 was another loco taken on by South West Trains and was used for the same purpose as 73201. The duties for a single loco for 'thunderbird' and driver training duties were proving a bit much. Here, we see the loco passing through Christchurch station hauling 3417 *Gordon Pettitt* on an eastbound driver training run from Bournemouth depot. 29 September 2005.

This is 73201 propelling 1497 *Freshwater* through Millbrook station heading towards Southampton Central on another driver training/unit turning run. The loco is unusually running on diesel at this point. 23 August 2005.

Here we see 73204 *Janice* leading 3-CEP 1198 and tailed by 73205 *Jeanette* and 73209 *Alison* passing through Eastleigh station, having originated from the works. The ensemble was going the short distance to the East Yard from where the three Class 73s were removed in favour of 66701 *Whitemoor*, which took the unit to Meldon Quarry in Devon of all places for storage. Of the locos, all are now Class 73/9s; 73204 has since been transformed into 73962, 73205 is now 73964 and 73209 has become 73961. 12 January 2005.

With 73207 in large logo blue and 73141 *Charlotte* in the short-lived 'Barbie' livery, the two GBRf operated locos are approaching Shawford hauling the rather dull 'mock teak'-liveried London Underground 4-TC unit as 5V38, the 09.20 Eastleigh Works to West Ruislip depot. The 4-TC unit was formerly used on the Waterloo to Weymouth services and has since been painted in a much more respectable maroon livery. 73207 has since been transformed into 73971 and based in Scotland, and 73141 is still in service with GBRf at the time of writing. 22 June 2012.

A nice blue combination of 73208 *Kirsten* and 4-VEP 3417 *Gordon Pettitt* are seen passing Totters Lane bridge, between Hook and Winchfield, as 5L23, the 12.50 Eastleigh Works to East Grinstead (Bluebell Railway). 73208 has since become 73965. 2 September 2010.

Above: 73235 is seen again, this time passing through platform four at Southampton Central hauling 4-VEP 3417 *Gordon Pettitt* and heading south for Bournemouth depot. 27 September 2005.

Left: Further south on the Waterloo to Weymouth line, this is 73201 again hauling 4-VEP 3417 *Gordon Pettitt* between Branksome and Parkstone, soon after leaving Bournemouth. The train went into the sidings at Poole before it headed north, probably as far as Eastleigh. 22 March 2005.

Having run round in Poole sidings, the loco is now hauling the train north. A second view of the same train is captured, this time as it passes Pokesdown station, just east of Bournemouth. The bright yellow flowers of the gorse bushes are just coming into full bloom with spring just around the corner. The wide space between the running lines here is due to there having been two centre roads through the station until they were removed in the late 1960s. 22 March 2005.

Most of these SWT runs just used one 73, as they could be used in push-pull mode with the EMU stock, and it was quite unusual to see one top and tailed with two locos. This, however, is 73235 at Branksome station about to take the short line to enter Bournemouth depot. 73109 *Battle of Britain* was on the other end of the unit and had led the train south from Eastleigh. 10 July 2008.

Until September 1995, Ashurst (New Forest) station was known as Lyndhurst Road, but it had always been much nearer to Ashurst than Lyndhurst, so for once it was a renaming that actually made a lot of sense! Having run round the unit, 73235 and 3417 *Gordon Pettitt* are seen again heading south with the return of the train seen earlier. 29 September 2005.

To end this series of images involving 'drags', we see 73235 once again as it approaches Millbrook hauling both 1498 *Farringford* and 3417 *Gordon Pettitt* with another training run from Bournemouth. Notice Freightliner's 57003 *Freightliner Evolution* in the background, which later moved to DRS but has just recently moved on again, this time to Locomotive Services and is currently awaiting further developments. 27 September 2005.

A few shots of loco convoys now. This is 73206 *Lisa*, 73208 *Kirsten* and 73136 *Perseverance* approaching Shawford station as 0Y66, the 09.20 Eastleigh Works to Selhurst depot. 73206 has now become 73963, 73208 is now 73965 and 73136 still gives sterling service for GBRf at the time of writing. 1 April 2010.

Another threesome as 73206 *Lisa*, 73208 *Kirsten* and 73141 *Charlotte* approach Eastleigh station heading for the nearby works. 8 September 2011.

A mixed selection of locos this time as we see 73212 *Shiney*, 73206 *Lisa*, 66710 and 66402 heading away from the camera as they pass through Tonbridge station making their way from Tonbridge Yard to Slade Green depot for maintenance. The name applied to 73212 at the time was unofficial, while the loco is currently still in service with GBRf. 73206 is now 73963, 66710 is in everyday service, and 66402 has since become one of just a couple of Class 66s that have been scrapped. It was released from DRS and taken on by GBRf and re-numbered to 66734 but it was unfortunately derailed in Scotland on the West Highland Line in 2012 while working a North Blyth to Fort William freight service. The loco ended up down an embankment in a very inaccessible location and had to be cut up on site, although all reusable parts were recovered. A replacement for the loco was eventually sourced from abroad and entered service during 2022 as the second 66734. 17 September 2009.

An impressive convoy this time consisting of 73201 *Broadlands*, 73963 *Janice*, 73136 *Mhairi*, 73141 *Charlotte* and 73212 *Fiona* running as 0Y68, the 09.39 Tonbridge West Yard to Eastleigh East Yard drifting through Winchester station. 1 September 2021.

Not a convoy but a fine collection of 73s viewed on the stabling point at Eastleigh, all in different liveries. To the left are 73206 *Lisa*, 73207 and 73141 *Charlotte*, while to the right are 73205 *Jeanette* and 73208 *Kirsten*. Since this shot was taken, it has now become much rarer to see 73s stabled at Eastleigh, most are based at Tonbridge and mostly used for RHTT and SITT trains. 24 November 2011.

In the first section of this volume, we saw 73/1s at work on the Swanage Railway, now we see a few of the various 73/2s that have visited the heritage line. Firstly, two former Southern Region stalwarts are seen approaching Corfe Castle with 73207 and 33202 *Dennis G. Robinson* working a Swanage to Norden train during the 2014 diesel gala. 9 May 2014.

A couple of days later, the same pair seen in the previous shot look superb as they come around the curve at Afflington, between Harmans Cross and Corfe Castle, with a Norden-bound train. 11 May 2014.

Just south of the location seen in the last image is the Woody Hyde campsite. 73207 and 33202 are seen once again, earlier in the day, as they pass by making for Norden with a train from Swanage. 11 May 2014.

Right: Three years earlier saw the visit of 73205 *Jeanette* and 73136 *Perseverance* to the line. 73205 had just received a repaint into InterCity livery and looked very smart. The clean duo is seen approaching Corfe Castle with a Swanage to Norden train. 6 May 2011.

Below: Almost looking like a pair of new locos, 73205 *Jeanette* and 73136 *Perseverance* are seen again, this time heading south at the aforementioned Woody Hyde with a Norden to Swanage service. 8 May 2011.

A final view of the pair as they pass over Corfe Viaduct and above the National Trust car park on the approach to Norden with a train from Swanage. 6 May 2011.

This is the first of three views of one of several charters to have visited the line using the connection to Network Rail. At the time of this image, the section of line between Norden and Worgret Junction was only being used very occasionally. 73208 *Kirsten* and 73206 *Lisa* are approaching what was once the boundary between Network Rail and the Swanage Railway at Catseye, near Norden, with 1Z97, the 08.02 London Waterloo to Swanage 'Purbeck Conqueror' charter organised by UK Railtours. 27 March 2010.

The train was far too long to remain at Swanage until it was time to return so 73212, which had been attached to the rear of the train throughout, was entrusted to haul the whole ensemble with the two locos still on the rear up to beyond Norden where it could stable out of the way of timetabled Swanage services. In this view, we see 73212 passing slowly by as it splutters its way through Harmans Cross station with the empty stock. 27 March 2010.

During its layover at Norden, 73208 *Kirsten* ran round the train to join 73212 and the pair are now seen awaiting departure from Swanage with 1Z98, the 16.05 Swanage to Waterloo return 'Purbeck Conqueror' charter. 73206 *Lisa* was now on the rear throughout. 27 March 2010.

The 2010 diesel gala saw a bit of a Class 73 festival at Swanage! This is 73208 *Kirsten*, 73141 *Charlotte*, 37264 and 73119 *Borough of Eastleigh* arriving at Corfe Castle with the 5Z73 Eastleigh Works to Swanage hauling 4-VEP 3417 *Gordon Pettitt* in readiness for the commencement of proceedings the following day. 6 May 2010.

At the conclusion of the 2010 gala, the returning convoy was somewhat different and is seen formed up near Norden with 73213, 73212, 56101 and 73136 *Perseverance* ready to haul 3417 *Gordon Pettitt* as 5Z73, the 17.50 Motala Ground Frame to Eastleigh Works. 11 May 2010.

After an early departure, the same train as in the previous shot is seen slowly approaching the main line in the wide cutting near Worgret Junction. This is near to the point that Swanage Railway jurisdiction ends and Network Rail's begins. In 2012, 56101 was exported to Hungary and now operates for Floyd Zrt. 11 May 2010.

Left: This is 73208 *Kirsten* and 73136 *Perseverance* coming across Corfe Common and approaching Townsend farm bridge with a Swanage to Norden service. Note the ex-BR Standard tank loco on the rear. 11 May 2007.

Below: A couple of interesting charters ran during 2019 utilising London Underground's 4-TC unit. This is 73201 *Broadlands* (with 73107 *Tracy* on the rear) approaching the A351 road bridge at Afflington with 1Z73, the 09.44 London Waterloo to Swanage 'The Swanage Sunday Special No. 1' organised by UK Railtours. The TC unit looks a little better in this livery compared to the 'fake teak' pictured earlier. 28 July 2019.

One loco was retained by Southern upon release from Gatwick Express for thunderbird duties similar to the ones used by SWT at the time. Immaculate 73202 *Dave Berry* is seen at Three Bridges while on driver training duties. It is wearing a unique livery which suited the loco rather well. 13 November 2008.

A few years later and 73202 *Dave Berry* is now in a livery to match the Southern EMUs, again looking rather smart. It is seen arriving beneath the semaphore signals at Littlehampton with 0Z74, the 08.31 route learning special from Hastings. 26 March 2014.

A second view of the loco at Littlehampton. It has been a very elusive loco to photograph, as it does not see a great deal of use, especially in recent years. It also appears to make a good landing pad for one of the local seagulls! 26 March 2014.

Next, we move on to a series of test train shots, which have been one of the main duties of the locos over the years. This first one sees 73201 *Broadlands* (with 73107 *Redhill 1844–1994* on the rear) approaching Micheldever station on the South Western Main Line between Basingstoke and Winchester with the 1Q07 Hither Green to Eastleigh via the Fawley branch. 16 April 2013.

Not strictly a test train as such, but these coaches are Network Rail-owned. This is 73212 and 73141 *Charlotte* passing Basingstoke with 5Y19, the 12.17 Tonbridge to Eastleigh, consisting of two former Gatwick Express GLVs modified as de-icing vehicles. 3 May 2013.

Having just visited Portsmouth Harbour, this is 73201 *Broadlands* coming round the curve on the approach to Fratton with an unidentified test train that will probably finish its circuit at Hither Green or Eastleigh Works later in the day. 73109 *Battle of Britain* was on the rear out of sight. 21 June 2013.

A recent clean of the footbridge windows at Eastleigh afforded this view of 73201 *Broadlands* (with 73107 *Redhill 1844–1994* on the rear) passing through the up fast line with 1Q85, the 06.08 Woking Down Reception to Woking via Waterloo and Southampton Down Loop test train. 14 August 2013.

Further east on the South Western Main Line at Wimbledon West Junction, we see 73207 coming off the line from Sutton with a 1Q13 test train bound for London. 17 October 2012.

On the rear of the train in the previous shot was immaculate 73205 *Jeanette*. At the time, it was very much a surprise when this loco was turned out in InterCity livery by GBRf. 17 October 2012.

The late rising winter sunshine has just managed to illuminate 73212 *Fiona* and 73141 *Charlotte* as they approach Southampton Central with the 1Q38 06.08 Woking to Woking test train. 5 January 2016.

Large logo blue 73207 makes a fine sight as it leads 1Q05, the 06.32 Eastleigh to Eastleigh via Weymouth and Fawley test train on the approach to Totton, just west of Southampton. 73201 *Broadlands* is bringing up the rear. 31 July 2014.

The train has changed direction in Southampton Up Loop and is now heading back west to take a trip down the freight only line to Fawley. 73201 *Broadlands* is now leading and is seen crossing the causeway and approaching Totton from the opposite direction than the previous image. 31 July 2014.

While the train was down the branch line at Fawley, I was able to take up this position on the Mountbatten Way bridge at Millbrook as 73207 approaches with the same train heading east towards Southampton Central. The lines furthest to the left of this shot lead to/from the Western Docks complex. 31 July 2014.

73212 *Fiona* and 73213 *Rhodalyn* have just arrived at Woking as 5Z20 from Woking Carriage Holding Sidings to Woking via London Waterloo. This was a special test train providing the exceptionally rare sight of a loco-hauled train in platform three at Woking, which is invariably only usually used by Class 455/456 units. The formation made a couple of trips to Waterloo and back to Woking during the morning. It is also now unusual for standard Class 73s to be used on a test train rather than the re-engineered 73/9s. 12 August 2021.

With 73213 *Rhodalyn* now leading, this is 5Z21, the 08.37 Woking to Woking via Waterloo seen speeding eastbound through West Byfleet. 12 August 2021.

The last collection of images from this section focuses on some of the general duties carried out by Class 73/2s over the last few years, including charter, stock and freight moves. This is 73207 and 73109 rapidly approaching Hook with 5Z31, the 07.11 Eastleigh to Acton Lane Reception Sidings hauling a rake of Mk.1 coaches to be used on a charter. 22 August 2014.

Viewed from high above Bincombe Tunnel, between Dorchester and Weymouth, we see 73204 *Janice* and 73141 *Charlotte* about to enter the tunnel with 1Z93, the 16.38 Weymouth to London Waterloo 'The Weymouth Seaside Special' organised by UK Railtours. Unfortunately, 73141 only got as far as Worting Junction (near Basingstoke) on the outward run and 73204 soldiered on solo hauling the dead 73 and the train to Weymouth. On the return run seen here, it is crawling along towards the end of the four-mile climb from the seaside town, some parts as steep as 1 in 50, but surprisingly it kept almost perfectly to time and arrived back at London just three minutes late! 73204 has since become 73962. In the background can be seen the major earthworks being undertaken in connection with the new Weymouth relief road that was opened a couple of years later. 1 August 2009.

Further east on the same line as the previous shot is another fierce climb between Poole and Parkstone. This is 73204 *Janice* and 73209 *Alison* (with 66708 on the rear) approaching Parkstone station. This was a private charter for GBRf employees, and as it was only formed of six coaches, it would not have given these two locos much of a problem. 22 August 2007.

This time we are at the well-known location of Campbell Road bridge at Eastleigh as 73212 in 'Barbie' livery comes down the centre road through the station with 1Z06, the 07.22 London Victoria to Weymouth 'Weymouth and the Waverley' charter organised by UK Railtours. 73204 *Janice* was again on the rear. 3 September 2010.

Getting a clear shot here is very tricky! 73208 *Kirsten* and 73136 *Perseverance* with 73209 *Alison* and 73204 *Janice* on the rear are approaching Clapham Junction with 1Z28, the 12.28 London Victoria to London Victoria private charter that travelled via Clapham Junction, Barnes, Virginia Water, Woking, Guildford, Redhill, Purley and Selhurst before returning to London. Freshly repainted 73208 was named at Victoria prior to departure of the train. 10 August 2006.

Due to the removal of the bridge that this image was taken from, it is no longer possible to get such a view as this. 73213 *Rhodalyn* and 73128 *OVS Bulleid CBE* pass Millbrook station as they head west with 1Z15, the 09.43 London Victoria to Swanage. This was the first day of a four-day tour of the UK to celebrate 15 years of operation by GBRf. 20118 and 20132 were on the rear of the train. 8 September 2016.

Above: Trains and planes at Eastleigh. 73205 *Jeanette* and 73208 *Kirsten* are stabled next to the station awaiting their next duties. This image could almost have been taken 20 odd years previous to this. 18 November 2011.

Left: Still at Eastleigh, we see our friend 73204 *Janice* again passing light engine through the station as viewed from a nearby multi-storey car park. 28 February 2008.

As mentioned previously, some of the main duties for the 73s from the early 2010s up to present time are the RHTT and SITT trains that operate on many third rail routes in the south. Kensington Olympia is the location for the next four images, as we see 73207 bringing up the rear of 3W90, the 04.30 Tonbridge West Yard to Tonbridge West Yard, which was routed via a great many lines around the southeast before going as far as Willesden Junction. 17 November 2014.

Class 73/2

Above: After a five-minute turn round at Willesden Junction, 73207 is seen heading back southbound through Kensington Olympia with 3W90. 73128 *OVS Bulleid CBE* was on the rear.
17 November 2014.

Right: Four years later than the previous couple of images, 3W90 is seen again, this time in the bright autumn sunshine as 73201 *Broadlands* brings up the rear of the train as it departs north towards Willesden Junction.
10 October 2018.

The train is seen returning back through the station with 73201 *Broadlands* now leading. 73136 *Mhairi* was bringing up the rear. The building just in view to the far left was once a Motorail terminal during the 1970s and 1980s.
10 October 2018.

73213 *Rhodalyn* is seen this time passing through Fleet station with 3W10, the 09.39 Eastleigh Works to Tonbridge West Yard transit move. The RHTT wagons had been at Eastleigh for tyre turning. 28 November 2016.

There are some locations, even in modern times, that 73s are seldom seen at. Bristol Parkway is one such place! The SITT train was introduced in the UK during 2012, and two vehicles of one of these new Network Rail snow trains worked to Bristol St Philip's Marsh depot for an unknown reason. The train was worked in top and tail mode by Network Rail's 57312 and GBRf's 73208 *Kirsten*. The return working is seen departing eastbound and running via Swindon and Basingstoke with the 73 on the rear, running as 5D66, the 11.44 St Philip's Marsh to Eastleigh. I don't think a Class 73 has appeared at this location since, certainly not a 73/1 anyway. 23 February 2012.

73213 *Rhodalyn* is seen again, this time with 73212 *Fiona* on the rear as they depart the depot exit line with 5Y73, the 09.40 Eastleigh to Southampton Western Docks hauling four Mk.1 coaches. 4 May 2016.

The driver has changed ends in Eastleigh station, and the train now makes a very smoky start with 73212 *Fiona* now leading. The locos had come 'cold' off the depot and 73213 was clearing its throat! I suspect the reason it was on diesel was due to only being a short trip and the need for the diesel engine in the docks complex. The coaches were later distributed by road to various new owners around the country. 4 May 2016.

Although this combination of loco and coaches could quite feasibly have been observed some years earlier in Virgin CrossCountry loco-hauled days, this is not quite what it seems! 73201 *Broadlands* races east through Farnborough with 5Z73, the 09.26 Selhurst to Selhurst via Eastleigh loop driver training run. 26 January 2010.

73205 *Jeanette* is this time seen approaching Havant from the Chichester/Brighton direction with an unidentified engineers' train bound for Eastleigh Yard. The lines to the left of the shot go to London Waterloo via Petersfield and Woking. 12 February 2009.

Quite a combination this time, as 73207, 73109, 73006 and 73005 are seen passing through Eastleigh station running as 5Y08, the 09.30 Eastleigh Works to Tonbridge West Yard. The two Class 73/0s were on their way for eventual conversion to 73/9s, with 73005 becoming 73966 and 73006 becoming 73967, both of which are now used on the Caledonian Sleeper trains in Scotland. 28 August 2014.

For a brief period in the mid-2010s, a pair of locos were rostered to work 6M94, the 11.09 Northfleet to Willesden Euroterminal in connection with the Crossrail project. This brought the welcome sight of 73s on freight and 73205 *Jeanette* and 73208 *Kirsten* are seen in charge passing Earl's Court, with the London Underground depot just glimpsed to the right of the picture. 5 November 2013.

To conclude this section on Class 73/2s, we see a one-off freight operated by no less than four locos. The pleasing pairing of BR blue 73201 *Broadlands* and large logo blue 73207 is exiting Southampton Tunnel and approaching Southampton Central with 6G27, the 04.20 Earlswood (Surrey) to Eastleigh Yard via Horsham, Netley and Southampton Up Goods Loop long-welded rail train. 73136 and 73107 were on the rear of the train but out of view in the tunnel. 7 March 2014.

Chapter 4
Class 73/9

By the mid-2010s, many of the existing Class 73s were becoming very tired, having put in many years of sterling service. As a result, GBRf decided to completely rebuild a batch of 73/2s by fitting them with a more powerful engine, and these were designated 73/9. I often wonder that, with such a substantial rebuild, why they were not reclassified Class 75? Anyway, 11 were eventually rebuilt, with 73961–965 being for general use and 73966–971 being dedicated to the Caledonian Sleeper operations in Scotland, and as such those six have no electric traction facility.

After a few teething troubles, all the locos now seem to have settled into being pretty reliable machines. Two locos, 73104 and 73211, were also rebuilt by Network Rail and re-engined using a pair of engines similar to the ones used in Class 220/221 diesel units operated by Avanti and CrossCountry. These were dubbed 'Ultra 73' and renumbered 73951/952. However, apart from having worked a few test trains, they do not seem to get much use at present.

The two Network Rail 73/9s have proved to be quite elusive since the first one was completed in 2014, and both are usually based at Derby RTC. This, however, is 73951 (the former 73104) in the works complex at Eastleigh wearing its yellow NR livery. Note the totally remodelled front end with the removal of all the jumper cables and the addition of new light clusters. 14 December 2015.

Photos of these locos on their former stamping ground in the south are few and far between. This is 73952 *Janis Kong* passing through Farnborough running as 0Z73, the 08.11 Eastleigh Works to Derby RTC. 18 September 2020.

This time we see both locos, 73951 *Malcolm Brinded* and 73952 *Janis Kong*, entering the sidings at Woking after working down from Derby as 0Z73, the 10.00 Derby RTC to Woking Up Yard, to recover a failed test train. 15 June 2017.

Moving on now to the GBRf 73/9s, this is the first of a series of shots of them working test trains, one of their current main duties in the south. This is 73961 *Alison* crossing from the down to the up line at St Denys so as to gain entry to Southampton Up Yard sidings with the regular 1Q53, the 10.35 Eastleigh Works to Eastleigh Works circuit. 31 August 2016.

With the trees just coming into leaf, this is 73965 approaching Basingstoke leading 1Q66, the 08.42 Woking to Woking via Fawley test train with 73961 *Alison* bringing up the rear. 18 April 2018.

With their larger 1,600hp diesel engines, the 73/9s are far more able to travel over non-electrified routes without the worry of overheating, as was the case with the original 600hp engines. After taking the previous image, I was able to get this shot as 73965 comes off the non-electrified route from Romsey at Eastleigh East Junction. The train was routed via Andover and was to visit Fawley before returning to Woking. 18 April 2018.

A year later than the previous image, and a wet morning at Southampton Central sees 73964 *Jeanette* passing through with 1Q66, the 06.07 Woking Up Yard to Woking Up Yard via Marchwood and Ludgershall test train. 24 April 2019.

A couple of hours on from taking the last shot, things had brightened up considerably, and the train is seen returning with 73962 *Dick Mabbutt* now leading on the approach to Southampton Central, having visited the two freight-only branch lines to Fawley and Ludgershall. 24 April 2019.

Above: Swiftly turning the camera around, we see 73964 *Jeanette* bringing up the rear of the train seen in the previous shot. It is passing the old signal box that closed way back in the mid-1980s but is rather surprisingly still extant. 24 April 2019.

Right: A Class 73 'off the juice' at Andover as we see 73963 *Janice* pausing in the platform with 1Q22, the 11.06 Woking Carriage Holding Sidings to Woking Carriage Holding Sidings via Fawley. 8 September 2021.

Some fabulous autumn colours are on show as 73962 approaches Salisbury Tunnel with 1Q66, the 10.07 Woking Carriage Holding Sidings to Woking Up Yard via Westbury. Also many miles from the nearest third rail, this was possibly the first time the rebuilt 73/9s had visited as far west as Salisbury and Westbury and was quite a novelty at the time. Even in 2022, any Class 73s are rare at these locations. 2 November 2016.

The regular monthly 1Q53, the 12.13 Eastleigh East Yard to Southampton Up Yard via Wimbledon depot test train is captured passing through Basingstoke hauled by 73961 *Alison* with 73962 *Dick Mabbutt* bringing up the rear. 17 February 2022.

This time 73963 *Janice* is seen approaching Southampton Airport Parkway with the 1Q53 11.38 Eastleigh Works to Eastleigh Works test train. 15 March 2017.

Taking the down fast line at Farnborough is 73964 *Jeanette* heading south with 1Q66, the 11.01 Woking Up Carriage Holding Sidings to London Waterloo via Oxford and Theale! This was a very unusual routing, but it really shows just how much more flexible the 73/9s are than the original locos. 23 March 2022.

The final test train view now. Coming across the Redbridge Causeway, just west of Southampton are 73961 *Alison* and 73965 with 1Q66, the 08.42 Woking to Woking test train. The crane in the foreground is a legacy of the Redbridge Wharf complex that used to be a very busy place in years gone by. 18 April 2018.

A few charter trains next as 73961 *Alison* and 73963 *Janice* are seen having just arrived at Yeovil Pen Mill with 1Z40, the 07.30 Stevenage to Weymouth 'The Sunny South Express', which the 73s worked throughout. This is a quite unique spot in the area which still retains semaphore signals. 14 July 2018.

This charter is certainly way off region for 73s as 73961 *Alison* and 73964 *Jeanette* are captured approaching Parson Street, soon after departing Bristol Temple Meads, with 1Z61, the 07.18 Ashford International to Weston-super-Mare GBRf staff special. 18 June 2016.

Class 73/9

Described in more detail in the first section of this book under the Class 73/1s, this is 73962 *Dick Mabbutt* and 73963 *Janice* bringing up the rear of 1Z73, the 08.00 London Waterloo to Paignton 'The Herd of Wildebeest' charter on the approach to Teignmouth. 16 July 2016.

A rare visit further north this time as 73963 *Janice* and 73962 *Dick Mabbutt* top and tail the London Underground 4-TC unit through Oxford as 1Z73, the 07.14 Paddington to Long Marston, operating in connection with Rail Live 2018 that was being held at the time. 20 June 2018.

This time 73962 *Dick Mabbutt* is heading east through Kensington Olympia with 5Z19, the 07.01 Eastleigh Works to London Victoria empty stock movement. 20189 and 20205 on the rear of the train later worked the 'Unknown London Wanderer' private wedding charter. 7 April 2018.

Ironically, I have yet to see the Caledonian Sleeper 73/9s in service on their dedicated duties. However, far more unusual is the sight of any of them in their former 'home' territory in the south. With refurbished running gear, this is 73966 (the former 73005) stabled in the holding sidings at Eastleigh, just like old times! Various examples of the Scottish contingent have visited the works for attention, but their appearance is still unusual. 24 June 2021.

Above: When the Scottish 73s do arrive in the south, they are invariably hauled by another loco rather than under their own power. This is 73970 (formerly 73117) being dragged through Basingstoke as 0Z73 Leicester to Eastleigh Works by 47727 *Edinburgh Castle/Caisteal Dhun Eideann*, which is also in Caledonian Sleeper livery. 12 September 2019.

Right: Slightly nearer to their Scottish home, this is 73969 being hauled through Doncaster by 92014 as 0Z92 Craigentinny to Doncaster Decoy. 24 February 2022.

As the two versions of GBRf Class 73/9s can very rarely be seen together, they are difficult to compare, but this chance encounter in Eastleigh holding sidings gives the opportunity, revealing that the front ends are very different in particular. This is 73966 being shunted around by 08683, passing 73962 *Dick Mabbutt* stabled alongside. There is also a low-flying pigeon getting in on the act! 24 June 2021.

A close up of the '08' and 73 that were originally built in 1959 and 1962, respectively. It is quite amazing that both are still in service after 60 odd years! 24 June 2021.

Possibly the first pairing of a 73/9 and a 73/1 as 73961 *Alison* and 73119 *Borough of Eastleigh* are seen crossing Corfe Common on the Swanage Railway with a Norden to Swanage service during the 2017 diesel gala. To date, this is the only Class 73/9 to have visited the Dorset heritage line. 5 May 2017.

The first two of the GBRf 73/9s, 73962 *Dick Mabbutt* and 73961, were initially based at Tonbridge for testing and staff training. The pair are seen with one of their first trips approaching Sevenoaks with 6Y73, the 08.54 Tonbridge West Yard to Tonbridge West Yard circular. 24 February 2015.

At the time, despite the conversion of at least half a dozen 73/9s, not many had ever worked revenue earning trains, forever seeming to do short solo test/crew training runs off Tonbridge Yard. In late 2015, however, 73963 had been in use on the Totton (near Southampton) based Railvac, but only operating at night. However, on this day, a 6G30 14.30 Totton Yard to Eastleigh Yard appeared in the system, so a chance to see one actually on a revenue earning service occurred. Here, the full train, including the Railvac in the middle, is led by immaculate 73963 (on diesel power) with 66744 *Crossrail* on the rear on the approach to Southampton Central. 2 October 2015.

Other trains that saw the first use of 73/9s were the RHTT services over various third rail electrified lines in the south. This is 73965 passing through Clapham Junction with 3W90, the 04.30 Tonbridge West Yard to Tonbridge West Yard via Willesden Junction. 28 October 2015.

Of all the new Class 800/801/802 units that were delivered to Eastleigh for commissioning during the late 2010s, all were hauled by GBRf Class 66s except this one. 73965 and 73963 *Janice* are approaching Eastleigh with the 6X80 Dollands Moor to Eastleigh hauling 802203 and 802204, eventually destined for TransPennine Express. 24 April 2019.

Bit of a strange one this. 73964 *Jeanette* is seen stabled at Westbury while engaged on route learning duties. It was assumed the 73/9 was used so as not to tie up a 66/7, which was required for other duties. 30 March 2021.